XTREME UNIVERSE

ICE GIANTS

AND MORE COLD CURIOSITIES

A&D Xtreme
BOLD HI-LO NONFICTION
An imprint of Abdo Publishing
abdobooks.com

ANN WILLIAMS

TAKE IT TO THE XTREME!

GET READY FOR AN EXTREME ADVENTURE! THE PAGES OF THIS BOOK WILL TAKE YOU INTO THE THRILLING WORLD OF ASTRONOMICAL WONDERS. WHEN YOU HAVE FINISHED READING THIS BOOK, TAKE THE XTREME CHALLENGE ON PAGE 45 ABOUT WHAT YOU'VE LEARNED!

ABDOBOOKS.COM
Published by Abdo Publishing, a division of ABDO, PO Box 398166, Minneapolis, Minnesota 55439.

Printed in the United States of America, North Mankato, MN.
102024
012025

THIS BOOK CONTAINS RECYCLED MATERIALS

Design: Kelly Doudna, Mighty Media, Inc.
Production: Mighty Media, Inc.
Editor: Katherine Chu

Cover Photograph: NASA/JPL-Caltech/B. Jónsson
Interior Photographs: ALMA (ESO/NAOJ/NRAO)/R. Sahai, pp. 22–23; ChameleonsEye, pp. 20–21; Enhanced image by Kevin M. Gill (CC-BY) based on images provided courtesy of NASA/JPL-Caltech/SwRI/MSSS., pp. 6–7; ESA and the Planck Collaboration, pp. 26–27; European Space Agency, NASA, pp. 28–29; Evans, Charles D, U.S. Fish and Wildlife Service, pp. 42–43; Marcio Jose Bastos Silva/Shutterstock Images, pp. 30–31; NASA, pp. 38–39; NASA, ESA and The Hubble Heritage Team STScI/AURA, p. 44; NASA, ESA, CSA, and M. Zamani (ESA/Webb); Science: F. Sun (Steward Observatory), Z. Smith (Open University), and the Ice Age ERS Team, pp. 4–5; NASA, ESA, CSA, STScI, pp. 12–13, 14–15; NASA, ESA, Mark Showalter (SETI Institute), Amy Simon (NASA-GSFC), Michael H. Wong (UC Berkeley), Andrew I. Hsu (UC Berkeley), pp. 8–9; NASA/GSFC/Arizona State University, p. 37; NASA/Johns Hopkins University Applied Physics Laboratory/Southwest Research Institute, pp. 32–33, 34–35; NASA/JPL-Caltech, pp. 10–11, 16–17; NASA/JPL-Caltech/B. Jónsson, pp. 1, 18–19; NASA/JPL/USGS, pp. 36–37; Natakorn Sapermsap/Shutterstock Images, pp. 24–25; NSIDC/Atsuhiro Muto, pp. 40–41; T. & R. Annan & Sons/Wikimedia Commons, p. 24
Design Elements: Arafat/Adobe Stock (header background); pixel/Adobe Stock (universe); Sergey Nivens/Shutterstock Images (header background)

LIBRARY OF CONGRESS CONTROL NUMBER: 2024938304
PUBLISHER'S CATALOGING-IN-PUBLICATION DATA
Names: Williams, Ann, author.
Title: Ice giants and more cold curiosities / by Ann Williams
Description: Minneapolis, Minnesota : ABDO Publishing, 2025 | Series: Xtreme universe | Includes online resources and index.
Identifiers: ISBN 9781098295066 (lib. bdg.) | ISBN 9798384915119 (ebook)
Subjects: LCSH: Giant stars--Juvenile literature. | Stars--Juvenile literature. | Universe--Juvenile literature. | Outer space--Exploration--Juvenile literature. | Astronomy--Juvenile literature.
Classification: DDC 523.112--dc23

CONTENTS

CHAPTER 1

COLD, COLDER, COLDEST

Some of the coldest ice found in space is in the Chamaeleon 1 gas cloud (*light blue wisps*). Ice in this region is minus 440 degrees Fahrenheit (-262°C).

Space is generally very, very cold. The average temperature in space is about minus 454 degrees Fahrenheit (–270°C). There are also many **extremely** cold objects in space. They include giant frozen planets. These ice giants are some of the coldest things in the universe!

CHAPTER 2

WHAT IS AN ICE GIANT?

An ice giant is a type of planet. It is mostly made of gases and water. Ice giants are located far away from a star. Because they have no heat source, their water freezes. Ice giants are basically huge balls of ice in space.

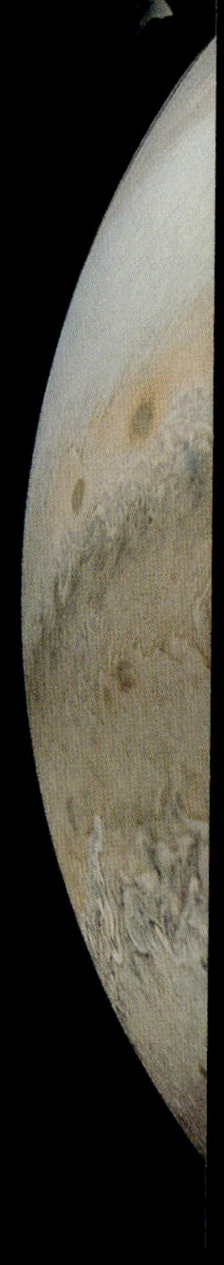

Ice giants have some of the same gases as gas giants such as Jupiter (*pictured*). These gases include hydrogen and helium.

Uranus (*left*) is minus 320 degrees Fahrenheit (-196°C). Neptune (*right*) is minus 330 degrees Fahrenheit (-201°C).

There are two ice giants in our solar system. These are Uranus and Neptune. They are the two planets farthest from the sun. Without the sun to warm them, these planets are very cold.

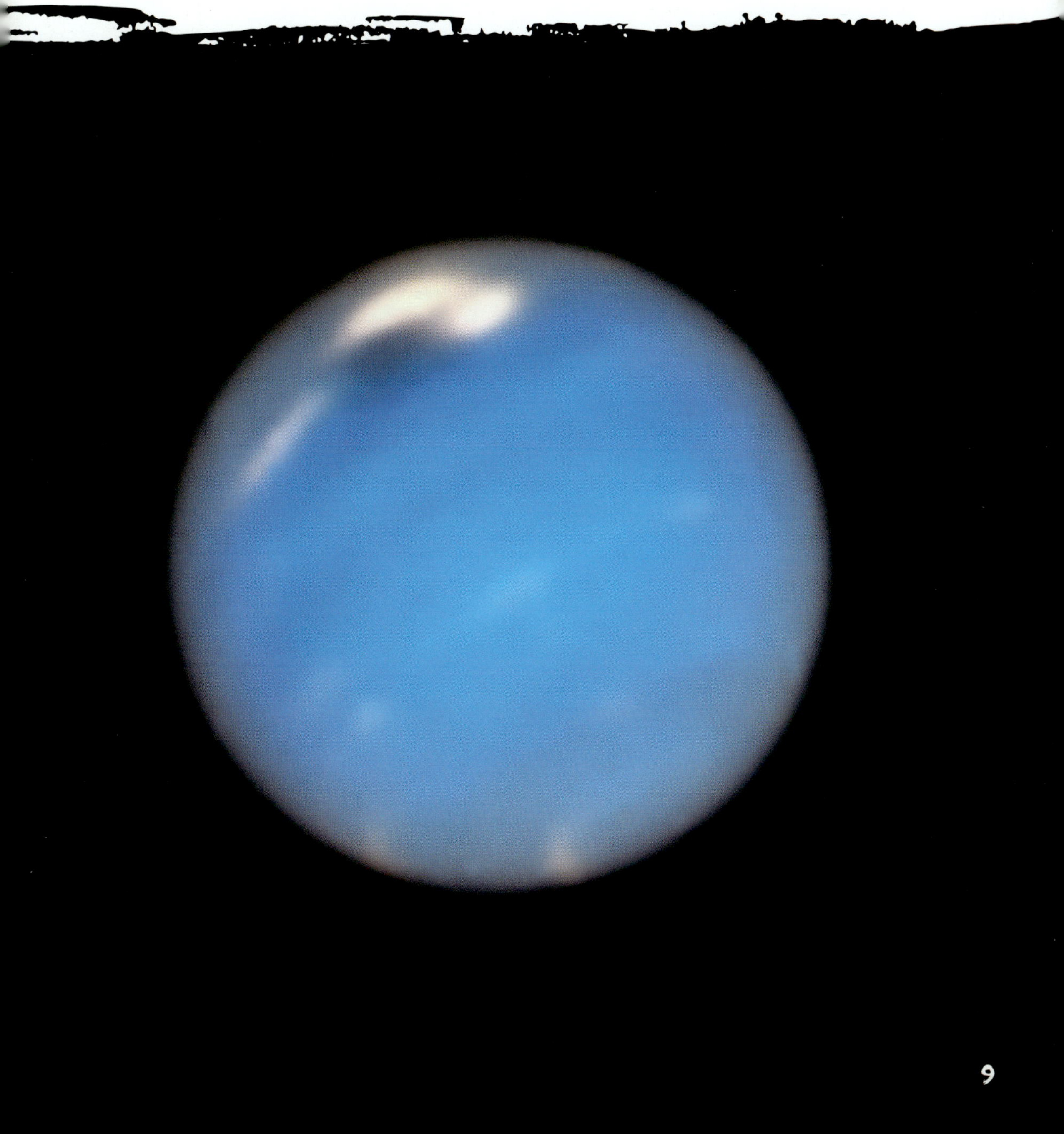

OGLE-2016-BLG-1195Lb is one of the coldest planets scientists have found outside the solar system, but it isn't an ice giant. It is 13,000 light-years away and about minus 400 degrees Fahrenheit (–240°C).

Scientists have only found one ice giant outside the solar system. They discovered it in 2014. The planet is 25,000 light-years away from Earth! Scientists are continuing to study this ice giant. They hope it will help them learn more about how Uranus and Neptune formed.

XTREME FACT

Size and distance in space are often measured in light-years. A light-year is the distance light travels in one year. One light-year equals about 5.88 trillion miles (9.46 trillion km).

CHAPTER 3

OBSERVING ICE GIANTS

Uranus was the first planet discovered by someone using a telescope. German-born British astronomer William Herschel first spotted it in 1781. However, he thought it was a **comet**. Herschel and other astronomers continued to observe and study Uranus. They soon realized that it is actually a planet.

An image of Uranus taken by the James Webb Space Telescope in 2023. It is possible to see Uranus with the naked eye if the sky is very clear and dark.

An image of Neptune taken by the James Webb Space Telescope in 2022. Le Verrier suggested naming the planet Neptune after the Roman god of the sea.

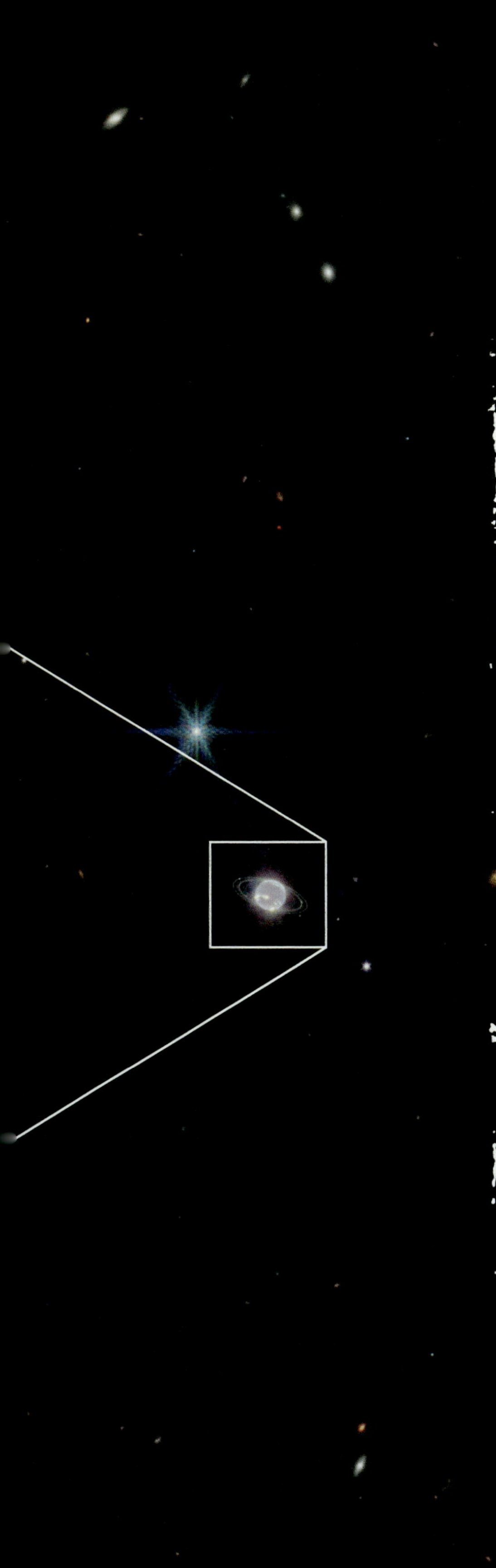

Neptune is too far away to be seen without a telescope. It was first observed in 1846 by German astronomer Johann Galle. He was able to find it with the help of French astronomer Urbain Le Verrier. Le Verrier had used a **mathematical theory** to **predict** where Neptune would be.

CHAPTER 4

VOYAGER PROGRAM

Humans can see Uranus and Neptune using telescopes. But the planets are so far away that it's hard to see them clearly. In 1977, the National Aeronautics and Space Administration (**NASA**) **launched** two space **probes**, Voyager 1 and Voyager 2. Their mission was to explore space. This meant they would also get a closer look at the planets in our solar system.

Almost 50 years after their launch, Voyager 1 (*pictured*) and Voyager 2 are still exploring space.

Voyager 2 passed Uranus in 1985 and Neptune in 1989. It sent photos and other information back to Earth. Voyager 2 found ten new moons orbiting Uranus. And it found six new moons orbiting Neptune. It also discovered that Neptune has four more rings than scientists previously thought.

Voyager 2's signals took about 2.5 hours to reach Earth during its observation of Uranus.

Voyager 2 flew by Neptune after having traveled 4.3 billion miles (7 billion km) from the time of its launch.

XTREME FACT

Voyager 2 recorded wind speeds of 450 miles per hour (724 kmh) in Uranus's atmosphere.

CHAPTER 5

THE BOOMERANG NEBULA

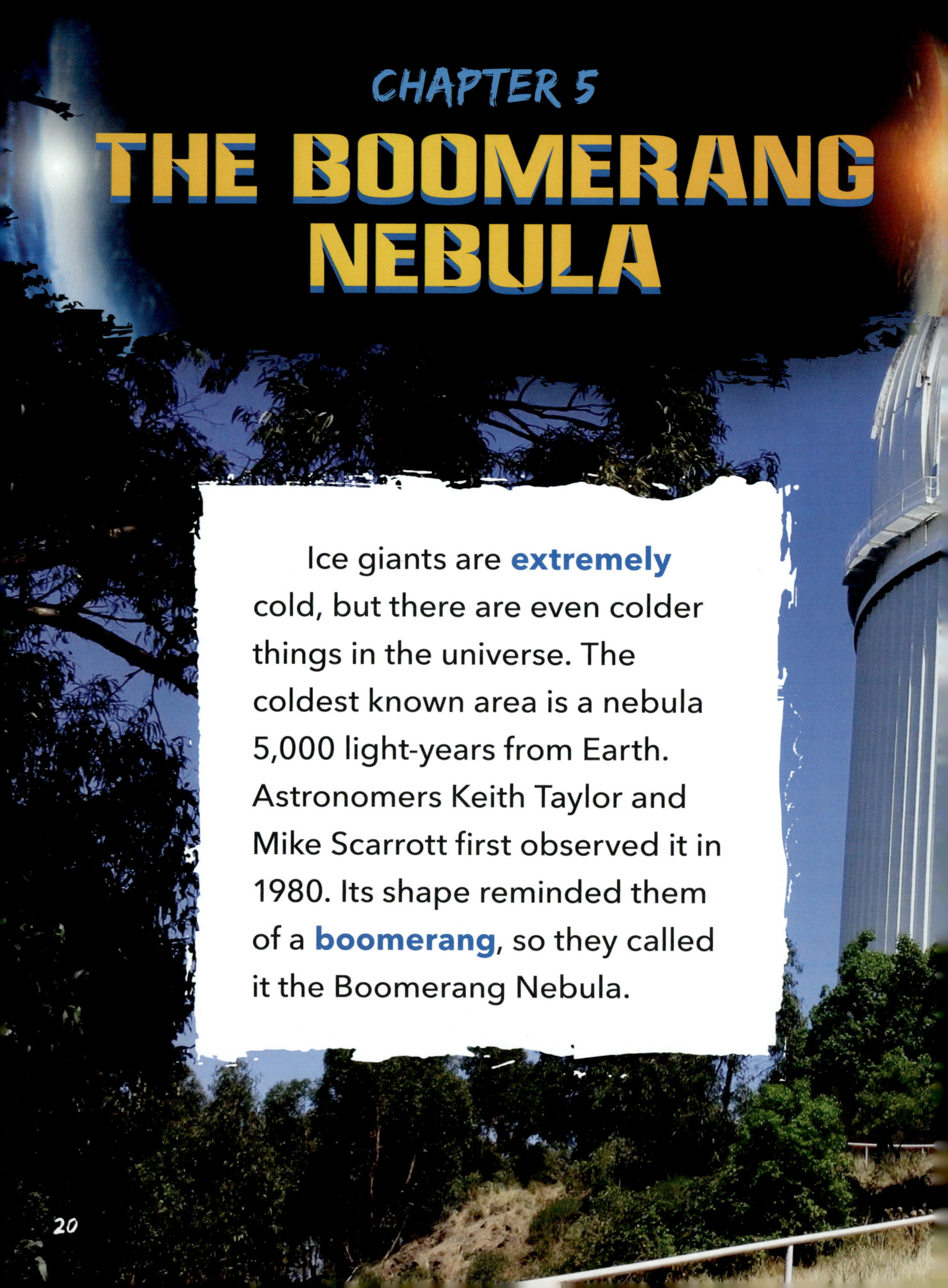

Ice giants are **extremely** cold, but there are even colder things in the universe. The coldest known area is a nebula 5,000 light-years from Earth. Astronomers Keith Taylor and Mike Scarrott first observed it in 1980. Its shape reminded them of a **boomerang**, so they called it the Boomerang Nebula.

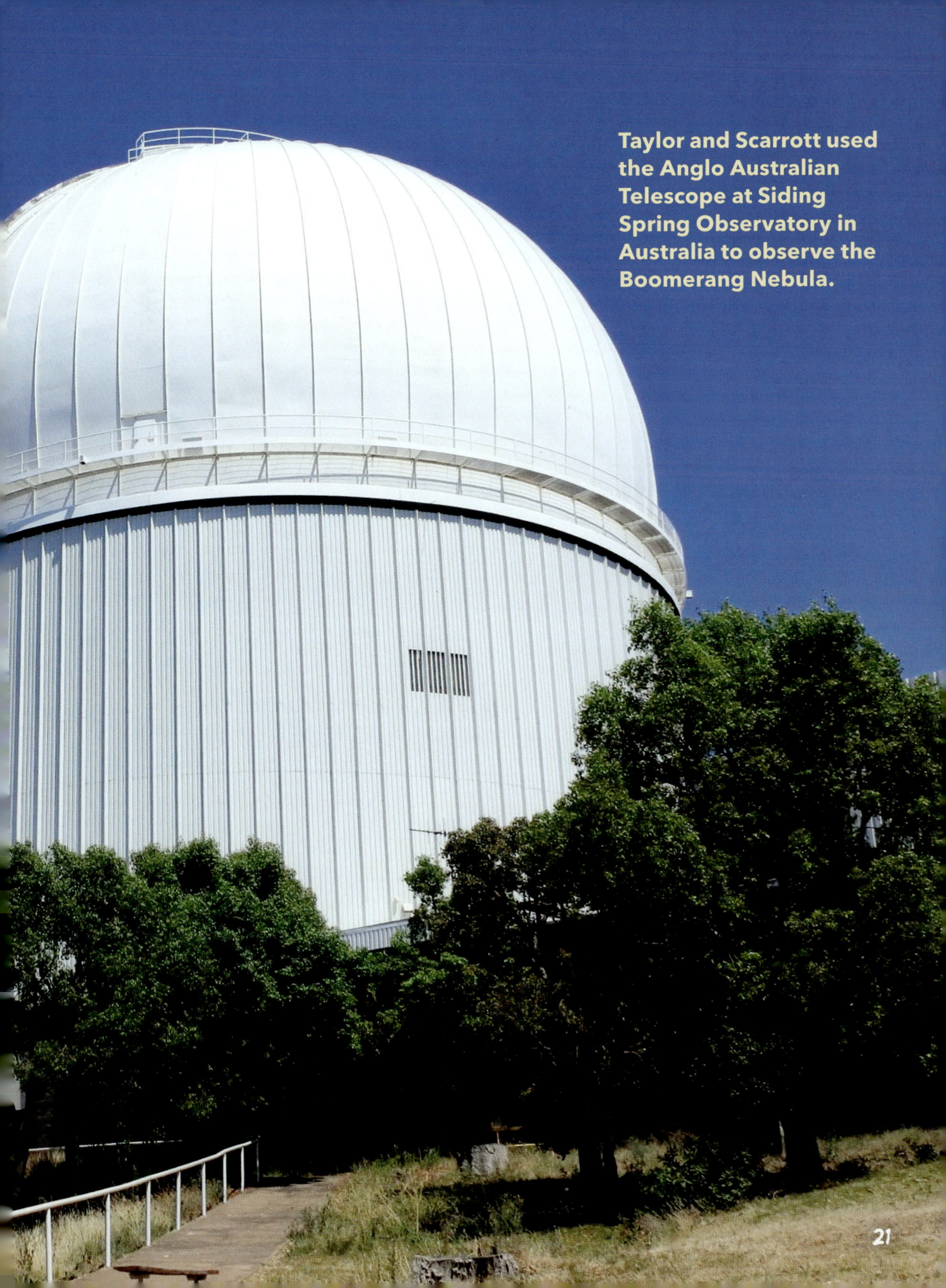

Taylor and Scarrott used the Anglo Australian Telescope at Siding Spring Observatory in Australia to observe the Boomerang Nebula.

The Boomerang Nebula can reach temperatures as low as minus 458 degrees Fahrenheit (-272°C).

A nebula is a large cloud of gas and dust in space. There are several kinds of nebulas. The **Boomerang** Nebula is a planetary nebula. This means it formed from gases released by a dying star. The temperature at the Boomerang Nebula is almost absolute zero. This makes it the coldest object in the universe.

XTREME FACT

Astronomers believe that in about 5 billion years, the sun will die and create a planetary nebula.

CHAPTER 6

KELVIN AND ABSOLUTE ZERO

In 1892, Thomson (*inset*) became Baron Kelvin. He chose the name Kelvin after a river in Glasgow, Scotland (*background*).

The most common scales for measuring temperature are Fahrenheit and Celsius. But there is a third scale called Kelvin. It is used mostly in science. British **mathematician** and scientist William Thomson proposed the Kelvin scale in 1848. In the Kelvin scale, 0 K is known as absolute zero.

All **normal** matter in the universe is made of atoms. Atoms are always moving and bumping into one another. This movement is what determines temperature. When atoms move faster, the matter they make up becomes warmer. When atoms move slower, the matter they make up becomes cooler. If atoms stop moving completely, the temperature of the matter would be absolute zero.

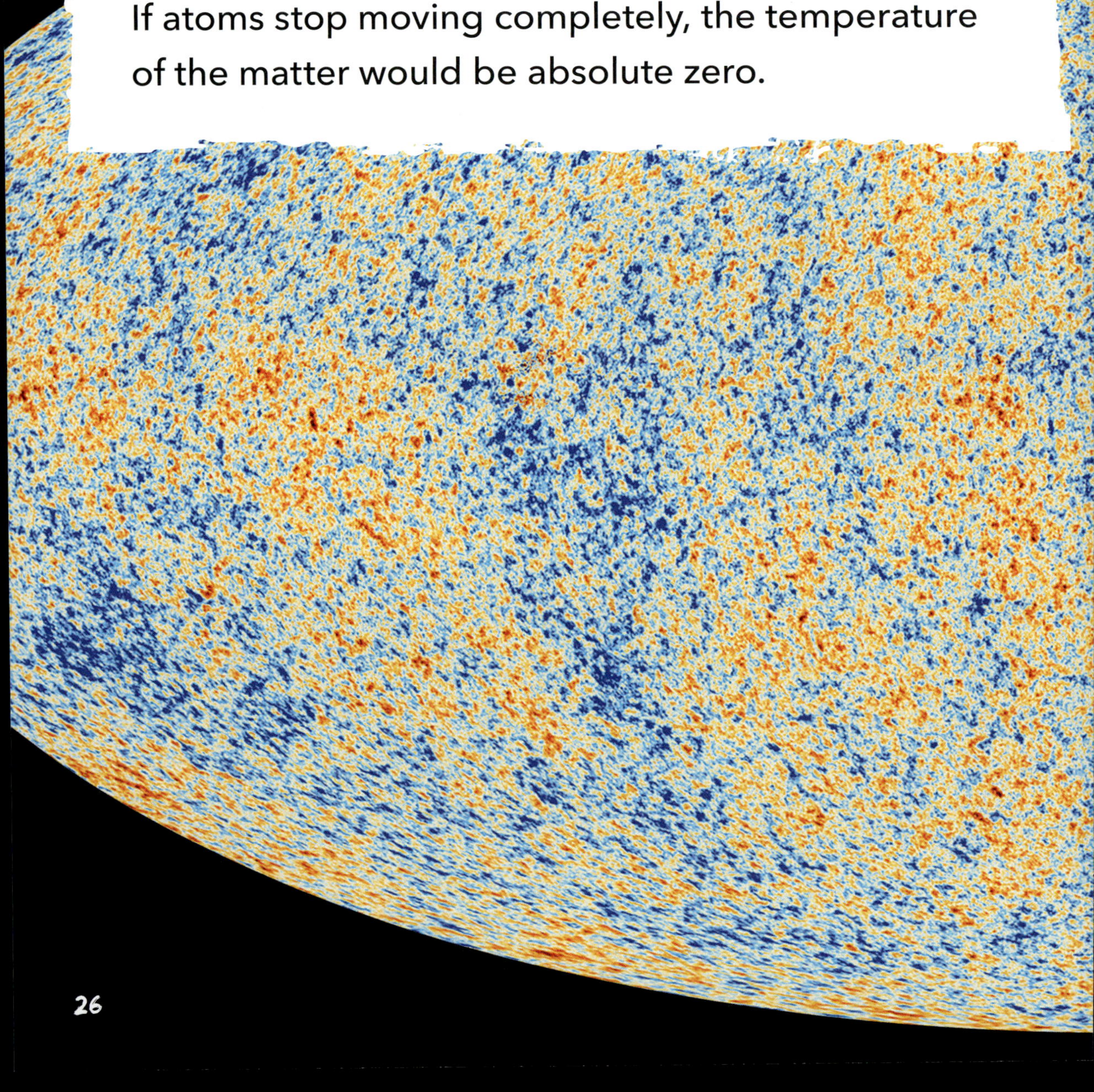

Differences in temperature (*blue and orange spots*) in different parts of the universe after it formed. Scientists believe this is when protons and electrons formed the first hydrogen atoms.

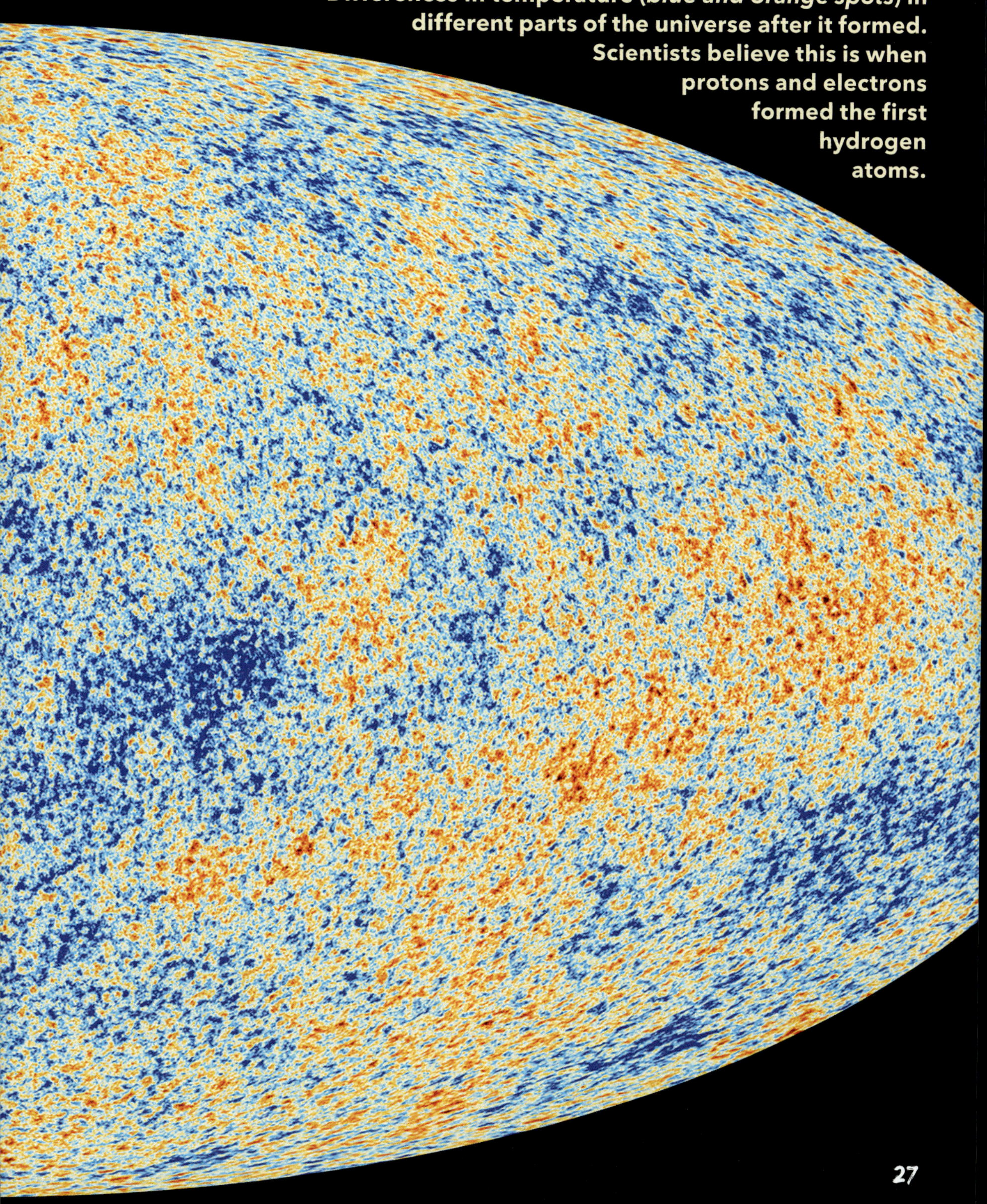

In 1995, astronomers Raghvendra Sahai and Lars-Åke Nyman observed and determined the temperature of the Boomerang Nebula.

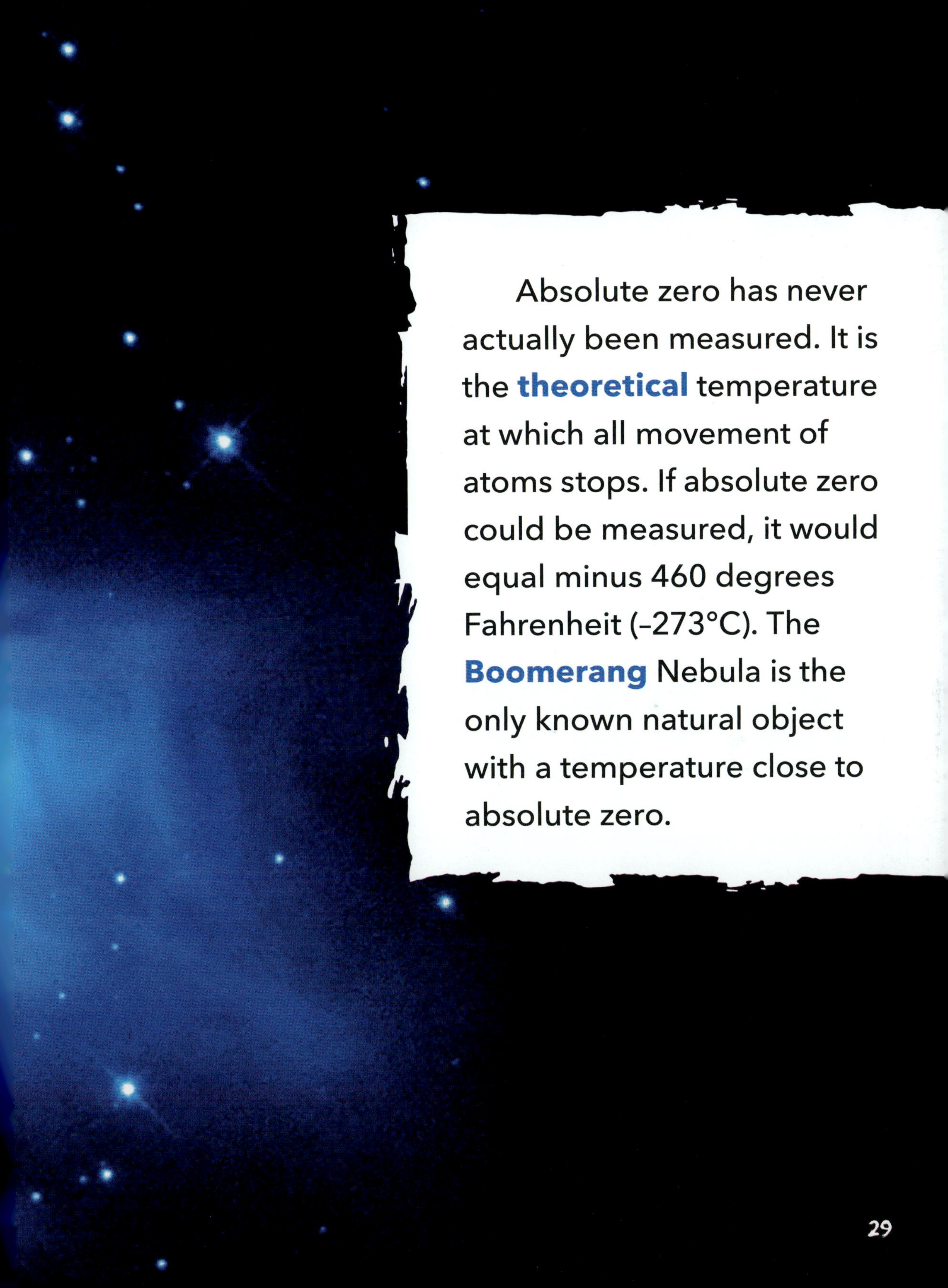

Absolute zero has never actually been measured. It is the **theoretical** temperature at which all movement of atoms stops. If absolute zero could be measured, it would equal minus 460 degrees Fahrenheit (–273°C). The **Boomerang** Nebula is the only known natural object with a temperature close to absolute zero.

Scientists have almost achieved absolute zero in laboratories. One of these labs is at the Massachusetts Institute of Technology (MIT) in Cambridge, Massachusetts. In 2003, scientists at the MIT lab recorded a temperature of 810 **trillionths** of a degree above absolute zero. This set the record for the lowest temperature ever measured!

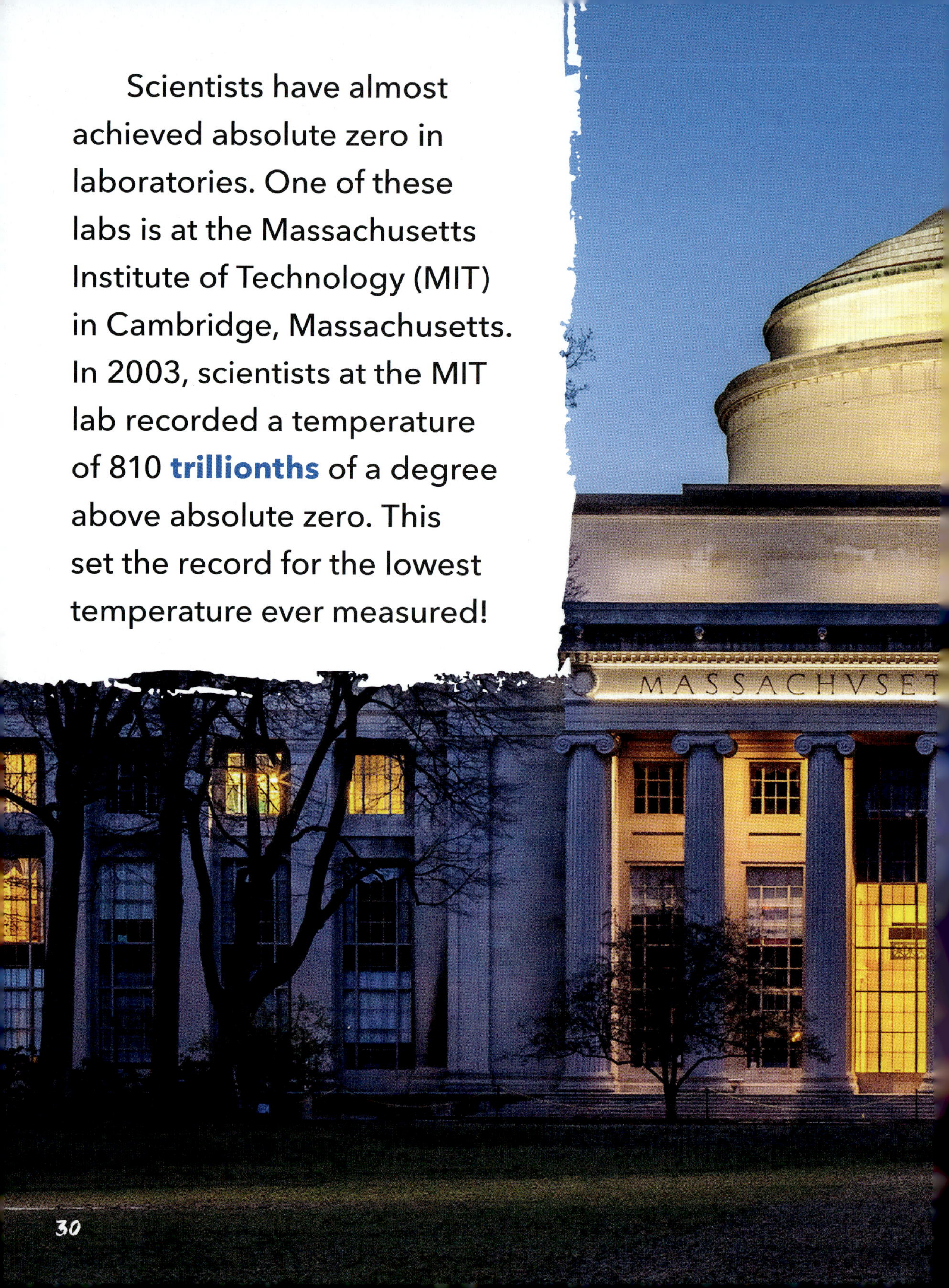

Scientists at MIT cooled sodium gas to reach the lowest recorded temperature.

CHAPTER 7

CHILLY NEIGHBORS

Besides ice giants, there are other very cold things in our solar system. One of them is the dwarf planet Pluto. It is one of the closer dwarf planets to the sun. Its surface temperature is minus 375 degrees Fahrenheit (–226°C).

Scientists named Pluto after the Roman god of the underworld

Pluto was discovered in 1930. From then until 2006, it was included as one of the planets in the solar system. In 2006, the International Astronomical Union changed the official definition of a planet. Pluto doesn't meet all the **criteria**. So, it is no longer considered one of the main planets.

Pluto (*below*) and Charon (*left*), the largest of its five moons

Earth's moon

Other cold places in our solar system can be found on parts of the moon. These include deep craters at the moon's north and south poles. Areas in these craters are very cold because the sun's light never reaches them. Inside the deepest craters, the temperature can be about minus 400 degrees Fahrenheit (–240°C).

The moon's Hermite Crater (*inset*) is the coldest natural place in the solar system. It is about minus 415 degrees Fahrenheit (–248°C).

CHAPTER 8

FREEZING EARTH

Earth also experiences some **extremely** cold temperatures. As Earth rotates and orbits the sun, the middle of the planet gets more direct sunlight. The sun's rays hit the north and south poles at more of an angle. So, the coldest places on Earth are generally near the north and south poles.

As of 2023, the average temperature on Earth's surface was 59 degrees Fahrenheit (15°C).

The coldest place on Earth is in Antarctica on the East Antarctic Plateau. Scientists there have recorded temperatures as low as minus 144 degrees Fahrenheit (–98°C). That is the lowest natural temperature ever recorded on Earth.

Scientists recorded the lowest temperatures in the East Antarctic Plateau (*pictured*) on a ridge between the summits of Dome Argus and Dome Fuji.

XTREME FACT

Even though it's covered with snow and ice, Antarctica is a desert. This is because it hardly ever rains there. It is the largest desert on Earth.

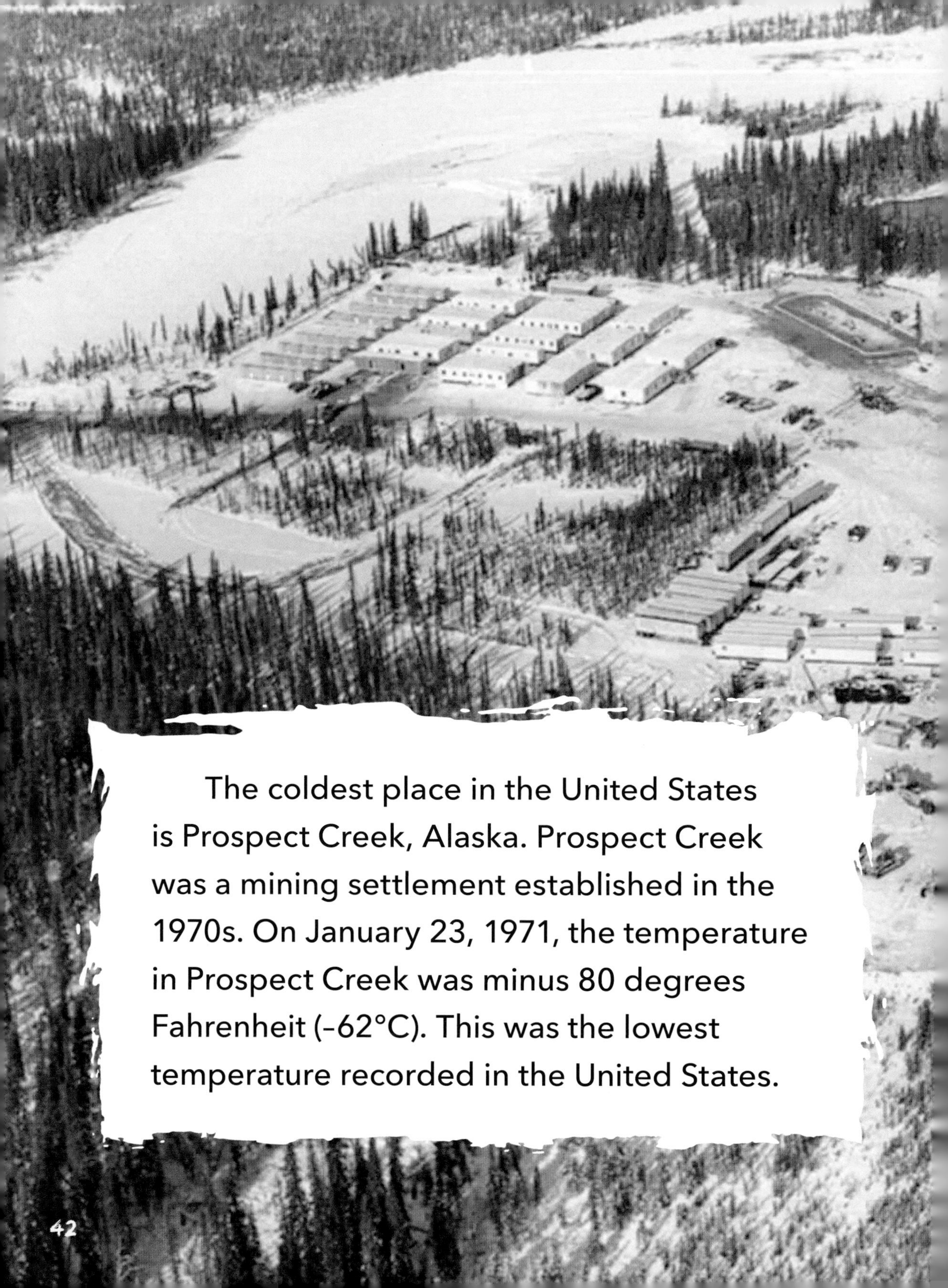

The coldest place in the United States is Prospect Creek, Alaska. Prospect Creek was a mining settlement established in the 1970s. On January 23, 1971, the temperature in Prospect Creek was minus 80 degrees Fahrenheit (–62°C). This was the lowest temperature recorded in the United States.

The warmest temperature Prospect Creek reached on January 23, 1971, was minus 64 degrees Fahrenheit (–53°C).

CHAPTER 9

COLDER THAN COLD

The coldest objects in the universe include ice giants, the **Boomerang** Nebula, and Pluto. The Voyager 1 and Voyager 2 **probes** are still exploring space. They continue to send information about the universe back to scientists on Earth. They may help find more ice giants!

The Boomerang Nebula

XTREME CHALLENGE

TAKE THE QUIZ BELOW AND PUT WHAT YOU'VE LEARNED TO THE TEST!

1) What is an ice giant?

2) Which planets in the solar system are ice giants?

3) What is the Boomerang Nebula?

4) What scale do scientists use to measure temperature?

5) Would you rather explore Uranus and Neptune or discover new ice giants?

GLOSSARY

billion–the number 1,000,000,000, or one thousand million.

boomerang–a curved flat object that when thrown correctly, returns to the thrower.

comet–a mass of ice and dust that moves through space and develops a tail as it nears the sun.

criteria–a rule or model that a decision is based on.

extremely–to a degree that exceeds the ordinary or expected.

launch–to send a spacecraft into space.

mathematical–relating to math, the science of numbers. A mathematician is someone who studies math.

NASA–National Aeronautics and Space Administration. NASA is a US government agency that manages the nation's space program and conducts flight research.

normal–most common.

predict–to guess something ahead of time on the basis of observation, experience, or reasoning.

probe—a spacecraft that attempts to gather information.

theoretical—relating to something based on theory rather than fact or experience.

theory—a general rule offered to explain a scientific event.

trillion—the number 1,000,000,000,000, or one thousand billion. One of one thousand billion equal parts is a trillionth.

ONLINE RESOURCES

Booklinks
NONFICTION NETWORK
FREE! ONLINE NONFICTION RESOURCES

To learn more about ice giants, please visit **abdobooklinks.com** or scan this QR code. These links are routinely monitored and updated to provide the most current information available.

INDEX